Information, History along with MYTHS IN PSYCHOLOGY by Dove Night

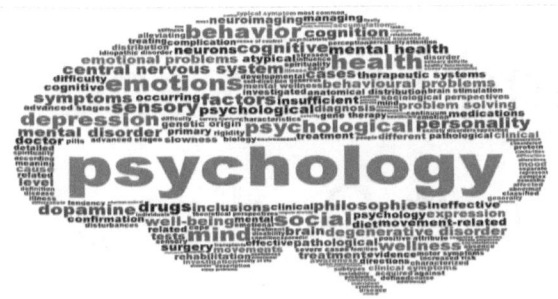

Table of contents

E. psychology and Greek mythology connected facts

Psychology today, the terrorism and mentally ill

A <u>General conversation about myths</u>

There are so many myths about psychology .I can tell you one of them ,psychology is not just about helping individuals with mental health issues , It is the study of the mental processes and behavior. The myths about psychology are endless. Some myths include that psychology studies lead to just mental health employment! The truth is, is that psychology is an academic discipline that involves scientific research like employment, relationships , adolescents and adults.

There were myths in psychology but the truth is psychology is a science that affects

everyone and
everything in all walks
of life.

B.<u>Some early history of modern psychology</u>

Modern psychology evolved largely from a German scientist named Wilhelm Wundt, he reportedly was one of the first to open a laboratory dedicated to psychological studies. He was one of the first experimental psychologists who focus on analyzing consciousness by the use of certain types of methods such as introspection and structuralism. Dr Wundt 's laboratory was the led to the development of many successful psychologists such as G.Stanley Hall and James mckeen Catell,who were became very famous psychologists in the fields of child psychology and Eugenics and mental tests.Wilheim Wundt was considered the first scientist to call himself a psychologist

Wundt with some of his colleagues in his laboratory

C.Dr Sigmund Freud and oedipal complex

Oedipus complex, a desire for sexual relations with the opposite sex partner and an associated sense of rivalry with the the same sex parent; a crucial stage in the developmental process. Sigmund Freud introduced the Oedipus theory in his *Interpretation of Dreams* book (1899). The term derives from the Greek mythology hero Oedipus who killed his father unknowingly and has intimate relations with his mother;

Freud applies the Oedipus complex to three to five year old children. He states that the phase

usually ended when the child becomes closer to the parent of the same sex
and <u>curbs</u> its sexual desires. If previous relationships with the parents and the parents' attitudes were loving and healthy, the stage is

passed through pleasantly. If trauma is presented, Dr Freud believes that, there will be an "infantile neurosis" that is an important predictor of infantile reactions when the child becomes an adult. The <u>superego</u>, which deals with the conscience, has its beginnings in the process of overcoming the Oedipus complex. Many psychologists later on however had found the Oedipus complex to be disproven and not

supported by empirical data.

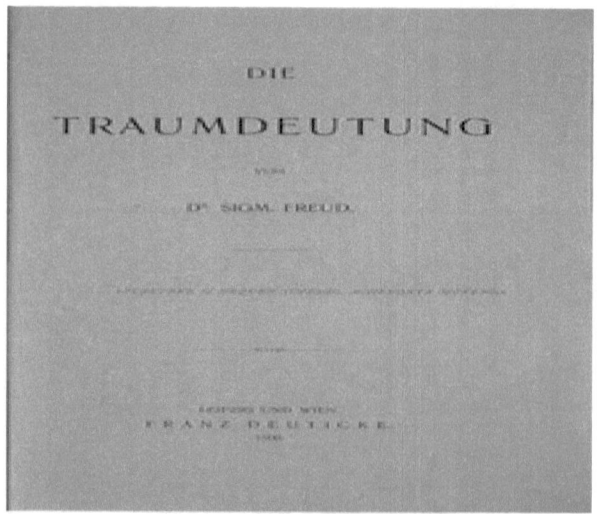

D.Dr Sigmund Freud's other achievements

SIGMUND FREUD
CONTRIBUTION'S TO
PSYCHOLOGY WERE
TREMENDOUS. He was
the founder of
psychoanalysis which
dealt with the

unconscious and dreams

Sigmund Freud was responsible for helping bring in the psychos sexual development theory which involves the seduction theory, the ID, ego and superego and theories related to dreams

Sigmund Freud greatly helped led to the beginnings of modern psychotherapy

E.Some Psychology and Greek mythology connected facts

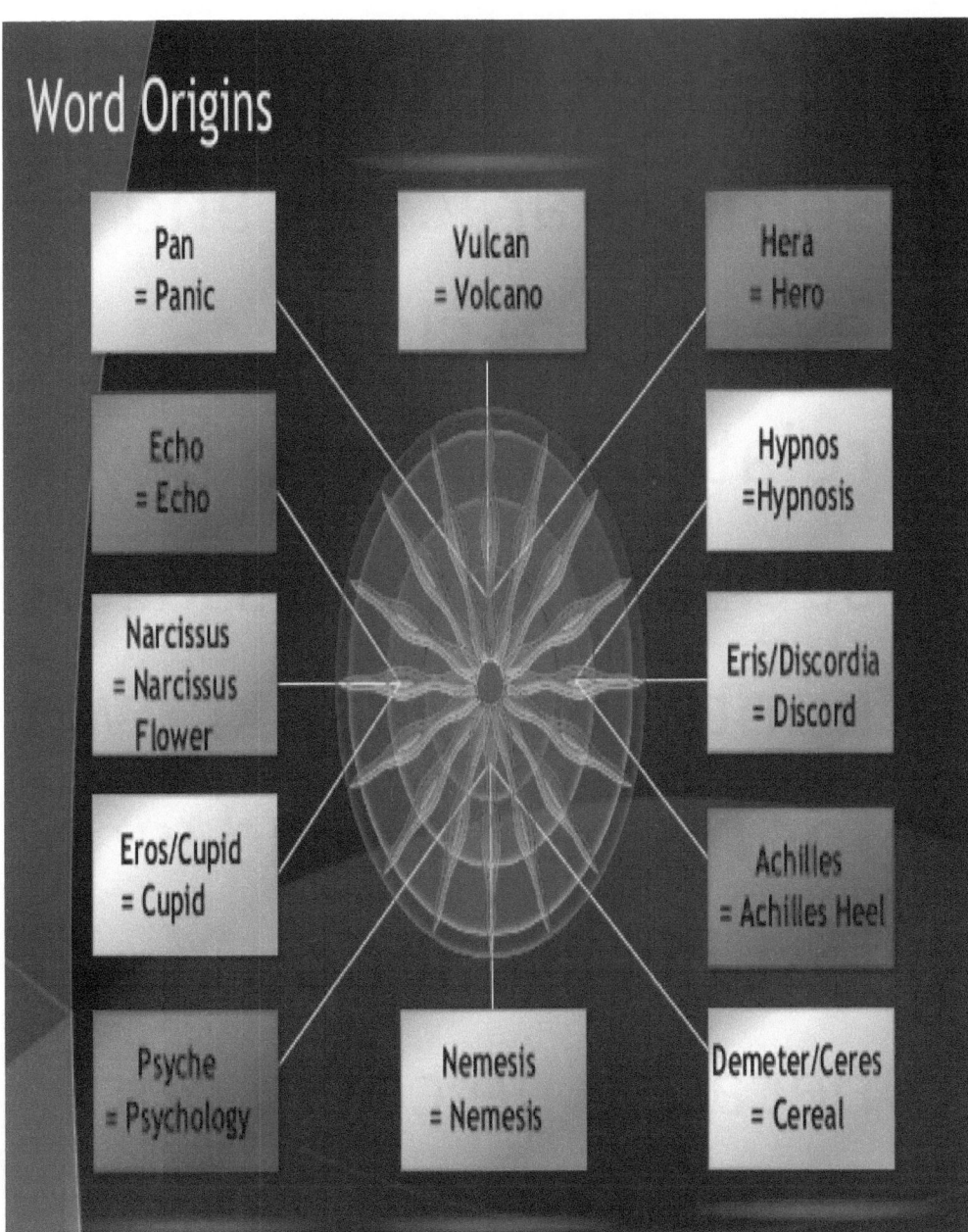

In Greek mythology, some characters and events are closely related to psychology such as the narcissus. A hunter who was known for his beauty and who lost his livelihood because he could not leave his reflection in a pool because he was infatuated with it. The term narcissm means someone who is fixated with their personal appearance and image. In psychology, one of the personality disorders/axis II disorders at play is narcissistic personality disorder which individuals cannot be cured but can receive and get therapy for their preoccupation with fame, success and beauty. These individuals can be callous and exploitative in their romantic relationships. These individuals have severe egocentrism and lack the ability to see that their problems and /or problems with others have to do with their egocentric actions. The treatment sometimes used for persons diagnosed with narcissistic personality disorder is

psychotherapy and medications.

Another word that represents a character or event in Greek mythology and psychology is panic. In Greek mythology, panic is the god of the wild,

hunting and companion
of the nymphs.

From Greek god pan,
the word panic disorder
is a

Anxiety disorder where individuals experience panic attacks that happen suddenly. It usually caused by stressful life situations, post-traumatic stress disorder and can be treated with psychotherapy and medications.

Psychology, terrorism and the mentally ill in America

JUNE 12, 2016

Remembering the lives lost in Orlando

Some horrific events such as The sandy hook massacre in December 2012,The Aurora Colorado tragedy in 2015 as well as the Orlando massacre in 2016 were committed by individuals with severe mental health problems and if they were treated and dealt with by law enforcement , mental health professionals , these tragedies would have probably been prevented. While most people diagnosed with mental health disorders are not violent. Unfortunately, a few are in accordance with the entire United States population. We must be careful and help those who are very sick and at the same time do what we can to help and protect others and prevent stigma which consists of staff helping their clients reach their full potential and help some mentally ill persons who needs to hospitalized or be on medication from committing senseless tragedies.